ARISTARCHUS OF SAMOS

On the Sizes and Distances of the Sun and Moon

Translated by

SIR THOMAS HEATH

ST. JOHN'S COLLEGE

(1913)

RCB CLASSIC REPRINT

Hidden Light Private Press – 2019

Rose City Books @ www.rosecitybooks.com

The Great Books of St. John's

ARISTARCHUS OF SAMOS
(c. 310 - 230 B.C.)

On the Sizes and Distances of the Sun and Moon

translated with notes by
SIR THOMAS HEATH

(Translation and Notes from *Part II* of *Aristarchus of Samos, the Ancient Copernicus* by *Sir Thomas Heath,* Oxford, *1913)*

This edition of 300 copies is published by arrangement with the Clarendon Press, Oxford, for the use of Members of St. John's College, Annapolis

Privately printed for members only
by
ST. JOHN'S COLLEGE
ANNAPOLIS

ARISTARCHUS ON THE SIZES AND DISTANCES OF THE SUN AND MOON

[HYPOTHESES]

1. *That the moon receives its light from the sun.*
2. *That the earth is in the relation of a point and centre to the sphere in which the moon moves.*[1]
3. *That, when the moon appears to us halved, the great circle which divides the dark and the bright portions of the moon is in the direction of our eye.*[2]
4. *That, when the moon appears to us halved, its distance from the sun is then less than a quadrant by one-thirtieth of a quadrant.*[3]
5. *That the breadth of the (earth's) shadow is (that) of two moons.*
6. *That the moon subtends one fifteenth part of a sign of the zodiac.*[4]

We are now in a position to prove the following propositions:—

1. *The distance of the sun from the earth is greater than eighteen times, but less than twenty times, the distance of the moon (from the earth)*; this follows from the hypothesis about the halved moon.

---

[1] Literally 'the sphere of the moon'.

[2] Literally '*verges* towards our eye', the word νεύειν meaning to 'verge' or 'incline'. What is meant is that the plane of the great circle in question passes through the observer's eye or, in other words, that his eye and the great circle are in one plane (cf. Aristarchus's own explanation in the enunciation of Prop. 5).

[3] I. e. is less than 90° by 1/30th of 90° or 3°, and is therefore equal to 87°.

[4] I. e. 1/15th of 30°, or 2°. Archimedes in his *Sand-reckoner* (Archimedes, ed. Heiberg, ii, p. 248, 19) says that Aristarchus 'discovered that the sun appeared to be about 1/720th part of the circle of the zodiac'; that is, Aristarchus discovered (evidently at a date later than that of our treatise) the much more correct value of $\frac{1}{2}°$ for the angular diameter of the sun or moon (for he maintained that both had the same angular diameter: cf. Prop. 8). Archimedes himself in the same place describes a rough method of observation by which he inferred that the diameter of the sun was less than 1/164th part, and greater than 1/200th part, of a right angle. Cf. pp. 311-12 *ante*.

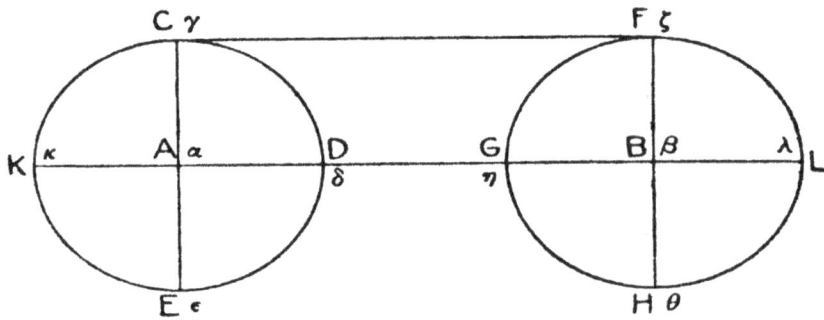

Fig. 16.

2. *The diameter of the sun has the same ratio (as aforesaid) to the diameter of the moon.*[1]

3. *The diameter of the sun has to the diameter of the earth a ratio greater than that which* 19 *has to* 3, *but less than that which* 43 *has to* 6; this follows from the ratio thus discovered between the distances, the hypothesis about the shadow, and the hypothesis that the moon subtends one fifteenth part of a sign of the zodiac.

PROPOSITION 1.

Two equal spheres are comprehended by one and the same cylinder, and two unequal spheres by one and the same cone which has its vertex in the direction of the lesser sphere; and the straight line drawn through the centres of the spheres is at right angles to each of the circles in which the surface of the cylinder, or of the cone, touches the spheres.

Let there be equal spheres, and let the points A, B be their centres.

Let AB be joined and produced;
let a plane be carried through AB; this plane will cut the spheres in great circles.[2]

Let the great circles be CDE, FGH.

Let CAE, FBH be drawn from A, B at right angles to AB; and let CF be joined.

---

[1] Pappus gives this second result immediately after the first result, i.e. before the parenthesis 'this follows from the hypothesis ...'. This arrangement seems at first sight more appropriate, and Nizze alters his text accordingly. But I think it better to follow the above order which is that of the MSS. and Wallis. One consideration which weighs with me is that the second result does not follow from the hypothesis of the halved moon alone; it depends on another assumption also, namely, that the sun and the moon have the same apparent angular diameter (see Prop. 8).

[2] Literally 'it will make, as sections in the spheres, great circles'; and then, in the next sentence, 'let it then make the circles CDE, FGH.' In translating these characteristic phrases, which occur very frequently, I wish I could have reproduced the Greek exactly, keeping the word 'sections', but it becomes impossible to do so when the phrase is extended so as to distinguish several sections made by one plane, e.g. one section in one sphere, one section in another sphere, and one section in a cone: Thus 'let it make, as sections, in the spheres, the circles CDE, FGH, and, in the cone, the triangle CEK' (Prop. 2) would be intolerable, with or without the multitude of commas, whereas clearness and conciseness is easily secured by saying 'let it cut the spheres in the circles CDE, FGH and the cone in the triangle CEK'.

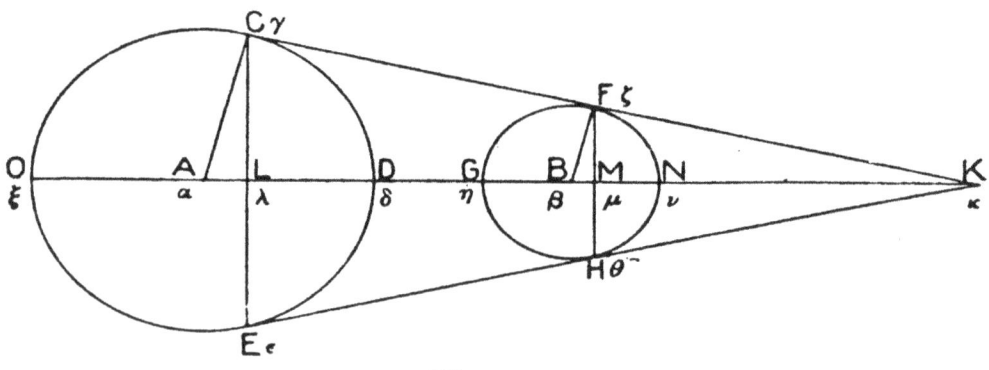

Fig. 17.

OF THE SUN AND MOON

Then, since CA, FB are equal and parallel, therefore CF, AB are also equal and parallel.

Therefore $CFAB$ is a parallelogram,
and the angles at C, F will be right;
so that CF touches the circles CDE, FGH.

If now, AB remaining fixed, the parallelogram AF and the semicircles KCD, GFL be carried round and again restored to the position from which they started, the semicircles KCD, GFL will move in coincidence with the spheres[1]; and the parallelogram AF will generate a cylinder, the bases of which will be the circles about CE, FH as diameters and at right angles to AB, because, throughout the whole motion, CE, HF remain at right angles to AB.

And it is manifest that the surface of the cylinder touches the spheres,
since CF, throughout the whole motion, touches the semicircles KCD, GFL.

Again, let the spheres be unequal, and let A, B be their centres; let that sphere be greater, the centre of which is A.

I say that the spheres are comprehended by one and the same cone which has its vertex in the direction of the lesser sphere.

Let AB be joined, and let a plane be carried through AB; this plane will cut the spheres in circles.

Let the circles be CDE, FGH;
therefore the circle CDE is greater than the circle GFH; so that the radius of the circle CDE is also greater than the radius of the circle FGH.

Now it is possible to take a point, as K (on AB produced), such that, as the radius of the circle CDE is to the radius of the circle FGH, so is AK to KB.

Let the point K be so taken, and let KF be drawn touching the circle FGH;
let FB be joined, and through A let AC be drawn parallel to BF;

---

[1] The force of κατά here is very difficult to render. The Greek phrase ἐνεχθήσεται κατὰ τῶν σφαιρῶν means 'will be carried, or move, *in* the spheres', that is, the circumferences of the semicircles will pass neither over nor under the surfaces of the spheres, but in coincidence with them throughout, in other words, they will by their revolution *describe* (as we say) the actual surfaces of the spheres.

358 ON THE SIZES AND DISTANCES

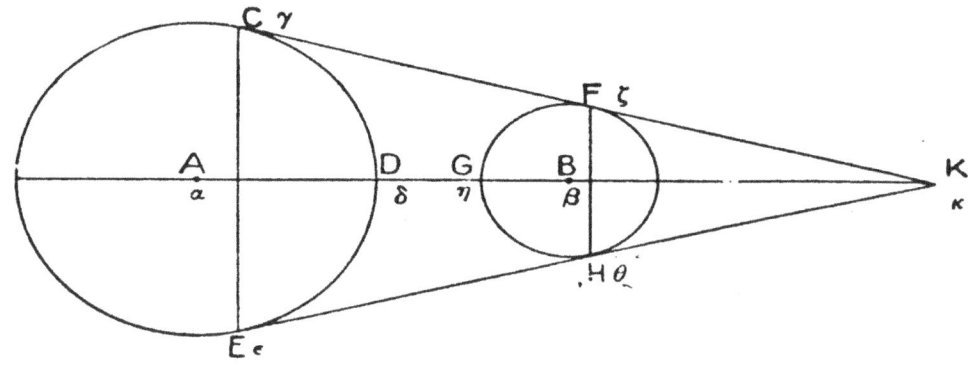

Fig. 18.

OF THE SUN AND MOON

let *CF* be joined.

Then since, as AK is to KB, so is AD to BN, while AD is equal to AC, and BN to BF, therefore, as AK is to KB, so is AC to BF.

And AC is parallel to BF; therefore CFK is a straight line.

Now the angle KFB is right; therefore the angle KCA is also right: therefore KC touches the circle CDE.

Let CL, FM be drawn perpendicular to AB.

If now, KO remaining fixed, the semicircles OCD, GFN and the triangles KCL, KFM be carried round and again restored to the position from which they started, the semicircles OCD, GFN will move in coincidence with the spheres; and the triangles KCL and KFM will generate cones, the bases of which are the circles about CE, FH as diameters and at right angles to the axis KL, the centres of the circles being L, M.

And the cone will touch the spheres along their surface, since KFC also touches the semicircles OCD, GFN throughout the whole motion.

PROPOSITION 2.

If a sphere be illuminated by a sphere greater than itself, the illuminated portion of the former sphere will be greater than a hemisphere.

For let a sphere the centre of which is B be illuminated by a sphere greater than itself the centre of which is A.

I say that the illuminated portion of the sphere the centre of which is B is greater than a hemisphere.

For, since two unequal spheres are comprehended by one and the same cone which has its vertex in the direction of the lesser sphere, [Prop. 1]
let the cone comprehending the spheres be (drawn), and let a plane be carried through the axis;
this plane will cut the spheres in circles and the cone in a triangle.

Let it cut the spheres in the circles CDE, FGH, and the cone in the triangle CEK.

It is then manifest that the segment of the sphere towards the circumference FGH, the base of which is the circle about FH as diameter, is the portion illuminated by the segment towards the circumference CDE, the base of which is the circle about CE as diameter and at right angles to the straight line AB;

for the circumference FGH is illuminated by the circumference CDE, since CF, EH are the extreme rays.[1]

And the centre B of the sphere is within the segment FGH;

so that the illuminated portion of the sphere is greater than a hemisphere.

PROPOSITION 3.

The circle in the moon which divides the dark and the bright portions is least when the cone comprehending both the sun and the moon has its vertex at our eye.

For let our eye be at A, and let B be the centre of the sun;

let C be the centre of the moon when the cone comprehending both the sun and the moon has its vertex at our eye, and, when this is not the case, let D be the centre.

It is then manifest that A, C, B are in a straight line.

Let a plane be carried through AB and the point D; this plane will cut the spheres in circles and the cones in straight lines.

Let the plane also cut the sphere on which the centre of the moon moves in the circle CD;

therefore A is the centre of this circle, for this is our hypothesis [Hypothesis 2].

Let the plane cut the sun in the circle EFR, and the moon, when the cone comprehending both the sun and the moon has its vertex at our eye, in the circle KHL and, when this is not the case, in the circle MNO;

and let it cut the cones in the straight lines EA, AG, QP, PR, the axes being AB, BP.

Then since, as the radius of the circle EFG is to the radius of the circle HKL, so is the radius of the circle EFG to the radius of the circle MNO,

---

[1] In Wallis's figure the letters F, H are interchanged. With his lettering, the extreme rays should be CH, EF. I have given F, H the positions necessary to suit the text, and my figure agrees with that of Vat.

ON THE SIZES AND DISTANCES

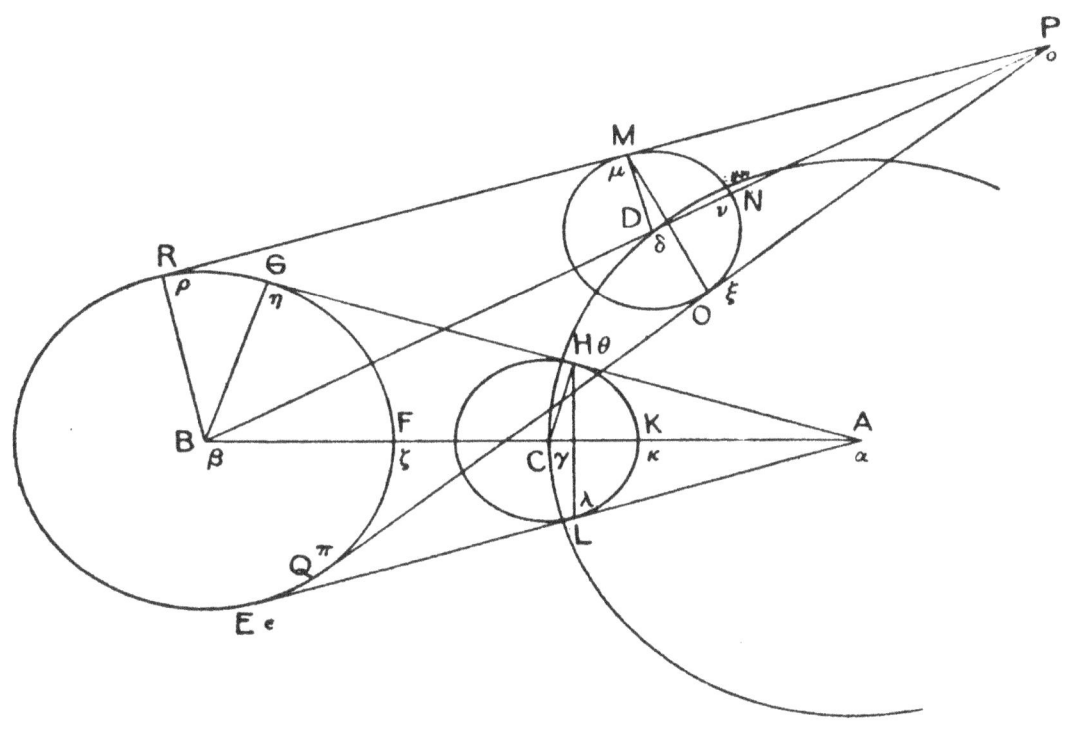

Fig. 19.

while, as the radius of the circle EFG is to the radius of the circle HLK, so is BA to AC,

and, as the radius of the circle EFG is to the radius of the circle MNO, so is BP to PD,

therefore, as BA is to AC, so is BP to PD,

and, *separando*, as BC is to CA, so is BD to DP;

therefore also, alternately, as BC is to BD, so is CA to DP.

And BC is less than BD, for A is the centre of the circle CD; therefore AC is also less than DP.

And the circle HKL is equal to the circle MNO;

therefore HL is also less than MO [by the Lemma[1]].

Accordingly the circle drawn about HL as diameter and at right angles to AB is also less than the circle drawn about MO as diameter and at right angles to BP.

But the circle drawn about HL as diameter and at right angles to AB is the circle which divides the dark and the bright portions in the moon when the cone comprehending both the sun and the moon has its vertex at our eye;

---

[1] The promised Lemma (the equivalent of which is stated, rather than proved, in Euclid's *Optics*, 24) does not appear. Some of the MSS. have a scholium containing a rather clumsy proof. A shorter proof is that of Nizze. We can use one circle instead of two equal circles; and we have to prove that, if A, P are points on the radius produced, P being further from the centre (C) than A

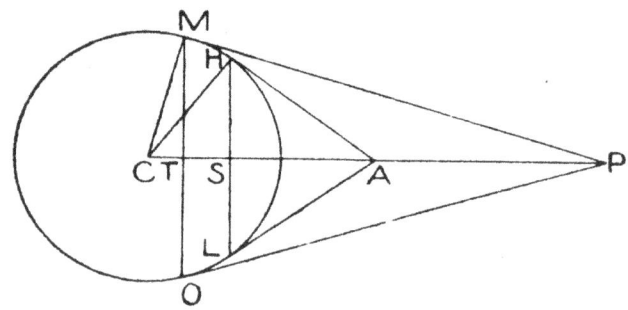

Fig. 20.

is, and if AH, AL be the pair of tangents from A, and PM, PO the pair of tangents from P, then $MO > HL$.

By Eucl. vi. 8 and 17, $CM^2 = CT.CP$, and $CH^2 = CS.CA$; therefore $CT.CP = CS.CA$, or $CA:CP = CT:CS$. But $CA < CP$; therefore $CT < CS$, so that the chord HL is less than the chord MO.

ON THE SIZES AND DISTANCES

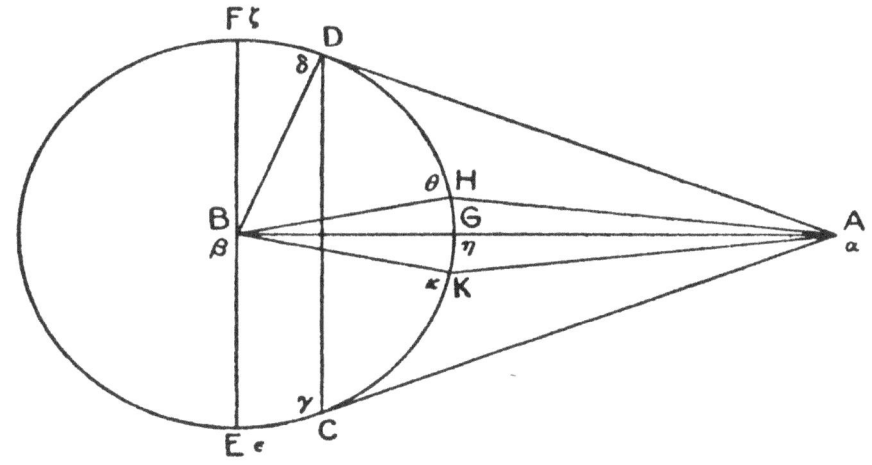

Fig. 21.

OF THE SUN AND MOON

and the circle about MO as diameter and at right angles to BP is the circle which divides the dark and the bright portions in the moon when the cone comprehending both the sun and the moon has not its vertex at our eye.

Accordingly the circle which divides the dark and the bright portions in the moon is less when the cone comprehending both the sun and the moon has its vertex at our eye.

PROPOSITION 4.

The circle which divides the dark and the bright portions in the moon is not perceptibly different from a great circle in the moon.

For let our eye be at A, and let B be the centre of the moon.

Let AB be joined, and let a plane be carried through AB; this plane will cut the sphere in a great circle.

Let it cut the sphere in the circle $ECDF$ and the cone in the straight lines AC, AD, DC.

Then the circle about CD as diameter and at right angles to AB is the circle which divides the dark and the bright portions in the moon.

I say that it is not perceptibly different from a great circle.

For let EF be drawn through B parallel to CD;
let GK, GH both be made (equal to) half of DF;
and let KB, BH, KA, AH, BD be joined.

Then since, by hypothesis, the moon subtends a fifteenth part of a sign of the zodiac,

366   ON THE SIZES AND DISTANCES

---

[1] This is a particular case of the more general proposition (similarly assumed by Archimedes in his *Sand-reckoner*) which amounts to the statement that, if each of the angles α, β is not greater than a right angle, and $\alpha > \beta$, then

$$\frac{\tan \alpha}{\tan \beta} > \frac{\alpha}{\beta}.$$

The proposition is easily proved geometrically (cf. Commandinus on the passage of the *Sand-reckoner*).

Let BC, BA make with ACD the angles α, β respectively, and let BD be perpendicular to AD.

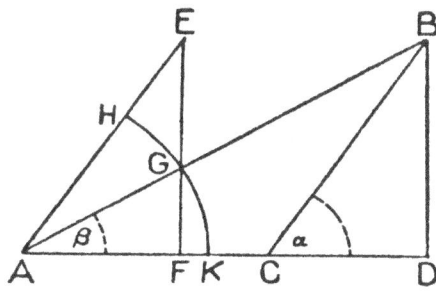

Fig. 22.

Now $\tan \alpha = BD/CD, \quad \tan \beta = BD/AD.$

We have therefore to prove that

$$AD : CD > \alpha : \beta.$$

OF THE SUN AND MOON

therefore the angle CAD stands on a fifteenth part of a sign.

But a fifteenth part of a sign is 1/180th of the whole circle of the zodiac,

so that the angle CAD stands on 1/180th of the whole circle; therefore the angle CAD is 1/180th of four right angles.

It follows that the angle CAD is 1/45th of a right angle.

And the angle BAD is half of the angle CAD;

therefore the angle BAD is 1/45th part of half a right angle.

Now, since the angle ADB is right,

the angle BAD has to half a right angle a ratio greater than that which BD has to DA.[1]

Accordingly BD is less than 1/45th part of DA.

Therefore BG is much less[2] than 1/45th part of BA, and, *separando*, BG is less than 1/44th part of GA.

Accordingly BH is also much less than 1/44th part of AH.

Cut off AF equal to CD, and draw FE at right angles to AD and equal to BD. Join AE.

Then $\angle EAF = \angle BCD = \alpha$.

Let EF meet AB in G.

Since $AE > AG > AF$, the circle with A as centre and AG as radius will cut AE in H and AF produced in K.

Now $\angle EAG : \angle GAF =$ (sector HAG) : (sector GAK)
$$< \triangle EAG : \triangle GAF$$
$$< EG : GF.$$

Componendo, $\angle EAF : \angle GAF < EF : GF.$

But $EF : GF = BD : GF = AD : AF = AD : CD.$

Therefore $\alpha : \beta < AD : CD,$

or $AD : CD > \alpha : \beta.$

In the particular application above made by Aristarchus $\alpha = \frac{1}{2}R$, so that $CD = BD.$

In this case therefore $AD : DB > \frac{1}{2}R : \angle BAD,$

or $BD : DA < \angle BAD : \frac{1}{2}R,$

that is to say, $\angle BAD : \frac{1}{2}R > BD : DA.$

[2] 'Much less', πολλῷ ἐλάσσων = 'less by much'. πολλῷ μείζων and πολλῷ ἐλάσσων are the traditional expressions used by Euclid and Greek geometers in general for 'a *fortiori* greater' and 'a *fortiori* less'. In Euclid the expressions have generally been translated 'much more then is ... greater, or less, than'. But there is no double comparative in the Greek. The idea is that, if a is, let us say, a *little* greater than b, and if c is greater than a, then c must be *much* greater than b.

---

[1] This is immediately deducible from a proposition given by Ptolemy (*Syntaxis*, I. 10, pp. 43–4, ed. Heiberg).

If two unequal chords are drawn in a circle, the greater has to the lesser a ratio less than the circumference (standing) on the greater chord has to the circumference (standing) on the lesser.

That is, if CB, BA be unequal chords in a circle, and $CB > BA$, then

(chord CB) : (chord BA) < (arc CB) : (arc BA).

Ptolemy's proof is as follows.

Bisect the angle ABC by the straight line BD, meeting the circle again at D. Join AEC, AD, CD

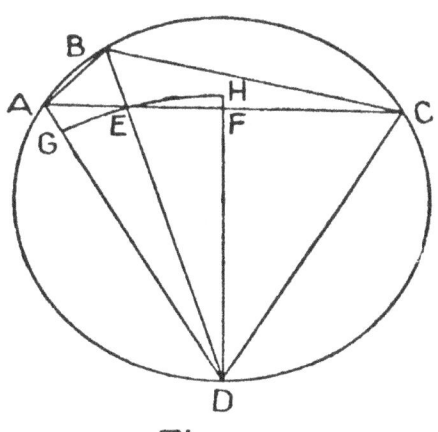

Fig. 23.

Then, since the angle ABC is bisected by BD,
$$CD = AD.$$ [Eucl., iii. 26, 29.]

And $$CE > EA.$$ [Eucl., vi. 3.]

Draw DF perpendicular to AEC.

OF THE SUN AND MOON

And *BH* has to *HA* a ratio greater than that which the angle *BAH* has to the angle *ABH*.[1]

Therefore the angle *BAH* is less than 1/44th part of the angle *ABH*.

And the angle *KAH* is double of the angle *BAH*, while the angle *KBH* is double of the angle *ABH*; therefore the angle *KAH* is also less than 1/44th part of the angle *KBH*.

But the angle *KBH* is equal to the angle *DBF*, that is, to the angle *CDB*, that is, to the angle *BAD*.

Therefore the angle *KAH* is less than 1/44th part of the angle *BAD*.

But the angle *BAD* is 1/45th part of half a right angle.

Accordingly the angle *KAH* is less than 1/3960th of a right angle.[2]

Now, since $DA > DE > DF$, the circle described with D as centre and DE as radius will cut AD between A and D, and will cut DF produced beyond F. Let the circle be drawn.

Since the triangle AED is greater than the sector DEG, and the triangle DEF is less than the sector DEH,

$$\triangle DEF : \triangle DEA < (\text{sector } DEH) : (\text{sector } DEG).$$

Therefore $FE : EA < \angle FDE : \angle EDA.$ [Eucl., vi. 1 and 33.]

Componendo, $FA : EA < \angle FDA : \angle EDA.$

Doubling the antecedents, we have

$$CA : AE < \angle CDA : \angle ADE,$$

and, *separando,* $CE : EA < \angle CDE : \angle EDA.$

But $CE : EA = CB : BA,$

and $\angle CDE : \angle EDA = (\text{arc } CB) : (\text{arc } BA).$

Therefore $CB : BA < (\text{arc } CB) : (\text{arc } BA).$

[The proposition is easily seen to be equivalent to the statement that, if α is an angle not greater than a right angle, and β another angle less than α, then

$$\frac{\sin \alpha}{\sin \beta} < \frac{\alpha}{\beta}.]$$

Now, since $\angle CDE = \angle CAB$ and $\angle ADE = \angle ACB$, in the same segments, we have

$$CB : BA < \angle CAB : \angle ACB,$$

or, inversely, $AB : BC > \angle ACB : \angle BAC,$

which is the property assumed by Aristarchus.

[2] $\frac{1}{2} \cdot \frac{1}{45} \cdot \frac{1}{44} = \frac{1}{3960}.$

But a magnitude seen under such an angle is imperceptible to our eye.

And the circumference KH is equal to the circumference DF; therefore still more is the circumference DF imperceptible to our eye;

for, if AF be joined, the angle FAD is less than the angle KAH.[1]

Therefore D will seem to be the same with F.

For the same reason, C will also seem to be the same with E.

Accordingly CD is not perceptibly different[2] from EF.

Therefore the circle which divides the dark and the bright portions in the moon is not perceptibly different from a great circle.

PROPOSITION 5.

When the moon appears to us halved, the great circle parallel to the circle which divides the dark and the bright portions in the moon is then in the direction of our eye; that is to say, the great circle parallel to the dividing circle and our eye are in one plane.

For since, when the moon is halved, the circle which divides the bright and the dark portions of the moon is in the direction of our eye [Hypothesis 3], while the great circle parallel to the dividing circle is indistinguishable from it,

therefore, when the moon appears to us halved, the great circle parallel to the dividing circle is then in the direction of our eye.

PROPOSITION 6.

The moon moves (in an orbit) lower than (that of) the sun, and, when it is halved, is distant less than a quadrant from the sun.

For let our eye be at A, and let B be the centre of the sun; let AB be joined and produced, and let a plane be carried through AB and the centre of the moon when halved;

this plane will cut in a great circle the sphere on which the centre of the sun moves.

---

[1] Pappus (pp. 560–8, ed Hultsch) gives an elaborate proof of this proposition depending on two lemmas; the proof, however, in the text as we have it, contains a serious flaw (p. 568. 2–3). But the truth of the assumption in Aristarchus's particular case is so obvious as scarcely to require proof.

[2] ἀνεπαίσθητος is strangely used with dat. as if equivalent to ἀνεπαισθήτως διάφορος or ἀδιάφορος πρὸς αἴσθησιν, 'imperceptibly different from'.

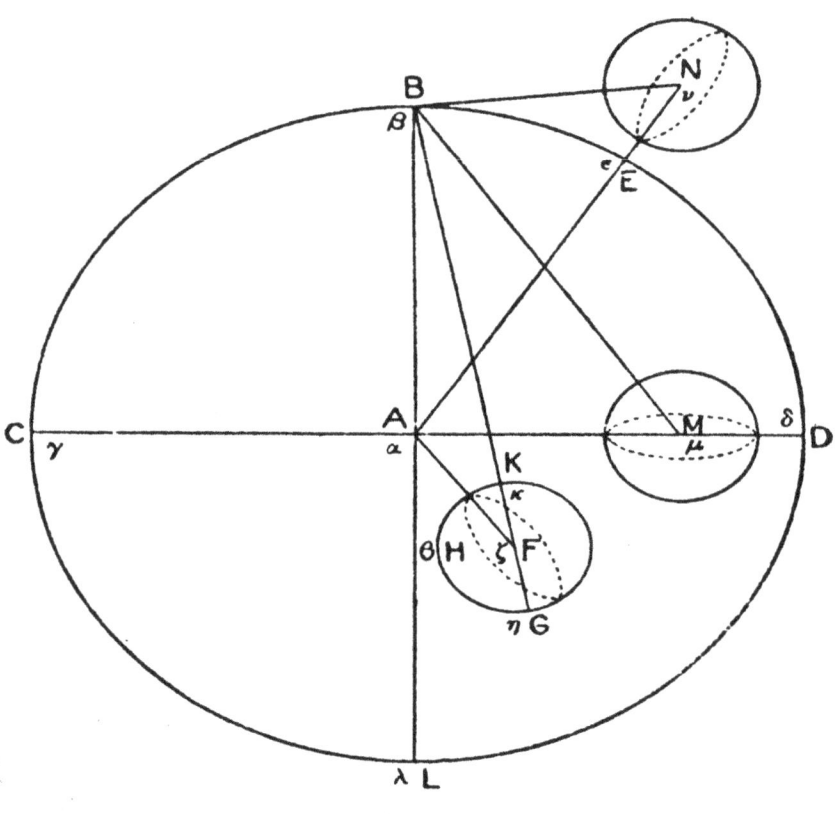

Fig. 24.

Let it cut it in the circle CBD; and from A let CAD be drawn at right angles to AB.

Then the circumference BD is that of a quadrant.

I say that the moon moves (in an orbit) lower than (that of) the sun, and, when halved, is distant less than a quadrant from the sun; that is to say, its centre is between the straight lines BA, AD and the circumference DEB.

For, if not, let its centre F be between the straight lines DA, AL, and let BF be joined;

then BF is the axis of the cone which comprehends both the sun and the moon,

and BF is at right angles to the great circle[1] which divides the dark and the bright portions in the moon.

Let the great circle in the moon parallel to the circle which divides the dark and the bright portions be GHK;[2] then since, when the moon is halved, the great circle parallel to the circle which divides the dark and the bright portions in the moon and our eye are in one plane [Prop. 5], let AF be joined.

Therefore AF is in the plane of the circle KGH.

---

[1] It is of course not actually a great circle, but a circle parallel to a great circle, which is however so close to it as to be indistinguishable from a great circle so far as our vision of it is concerned [Prop. 4]. The expression is therefore excusable, as in Hypothesis 3; there is no need to omit μέγιστον from the text as Nizze does.

[2] I have drawn the circle GHK and the other circles representing the sections of the moon as they are drawn in Wallis's figures; but I think the circles in the moon defining the dark and bright portions and, by hypothesis, in the same plane with our eye would be better represented by the dotted circles which I have added to the figure.

And BF is at right angles to the circle KHG, and therefore to AF; therefore the angle BFA is right.

But the angle BAF is also obtuse: which is impossible.

Therefore the point F is not in the space bounded by the angle DAL.

I say that neither is it on AD.

For, if possible, let it be M; and again let BM be joined, and let the great circle parallel to the dividing circle be taken, its centre being M.

Then, in the same way as before, it can be proved that the angle BMA [made with the great circle]¹ is right.

But the angle BAM is so also: which is impossible.

Therefore the centre of the moon, when halved, is not on AD.

Therefore it is between AB and AD.

Again, I say that it is also within the circumference BD.

For, if possible, let it be outside, at N; and let the same construction be made.

It can then be proved that the angle BNA is right; therefore BA is greater than AN.

But BA is equal to AE; therefore AE is also greater than AN: which is impossible.

Therefore the centre of the moon, when halved, will not be outside the circumference BED.

Similarly it can be proved that neither will it be on the circumference BED itself.

Therefore it will be within.

Therefore, &c.

---

words πρὸς τὸν μέγιστον κύκλον are in fact not wanted, and, if they are retained, cannot be taken with ὀρθή in the sense of '*at right angles* to the great circle'; they can only be taken closely with γωνία and as meaning 'towards the great circle', or 'made with the great circle'. But, as the words do not occur in the corresponding passage about the angle BNA lower down, I think they should be struck out, as an interpolation by some one who thought the inference wanted some further explanation but failed to supply it intelligibly.

PROPOSITION 7.

The distance of the sun from the earth is greater than eighteen times, but less than twenty times, the distance of the moon from the earth.

For let A be the centre of the sun, B that of the earth.

Let AB be joined and produced.

Let C be the centre of the moon when halved;
let a plane be carried through AB and C, and let the section made by it in the sphere on which the centre of the sun moves be the great circle ADE.

Let AC, CB be joined, and let BC be produced to D.

Then, because the point C is the centre of the moon when halved, the angle ACB will be right.

Let BE be drawn from B at right angles to BA;
then the circumference ED will be one-thirtieth of the circumference EDA;
for, by hypothesis, when the moon appears to us halved, its distance from the sun is less than a quadrant by one-thirtieth of a quadrant [Hypothesis 4].

Thus the angle EBC is also one-thirtieth of a right angle.

Let the parallelogram AE be completed, and let BF be joined.

Then the angle FBE will be half a right angle.

Let the angle FBE be bisected by the straight line BG; therefore the angle GBE is one fourth part of a right angle.

But the angle DBE is also one thirtieth part of a right angle; therefore the ratio of the angle GBE to the angle DBE is that which 15 has to 2:
for, if a right angle be regarded as divided into 60 equal parts, the angle GBE contains 15 of such parts, and the angle DBE contains 2.

Now, since GE has to EH a ratio greater than that which the angle GBE has to the angle DBE,[1]

---

[1] The proposition assumed is again the equivalent of the fact that $\frac{\tan \alpha}{\tan \beta} > \frac{\alpha}{\beta}$, where each of the angles α, β is not greater than a right angle and $\alpha > \beta$. (Cf. note on pp. 366-7, above.) Let the angles α, β be the angles GBE, HBE respectively in the subjoined figure (Fig. 26). Let GE be perpendicular to BE

378 ON THE SIZES AND DISTANCES

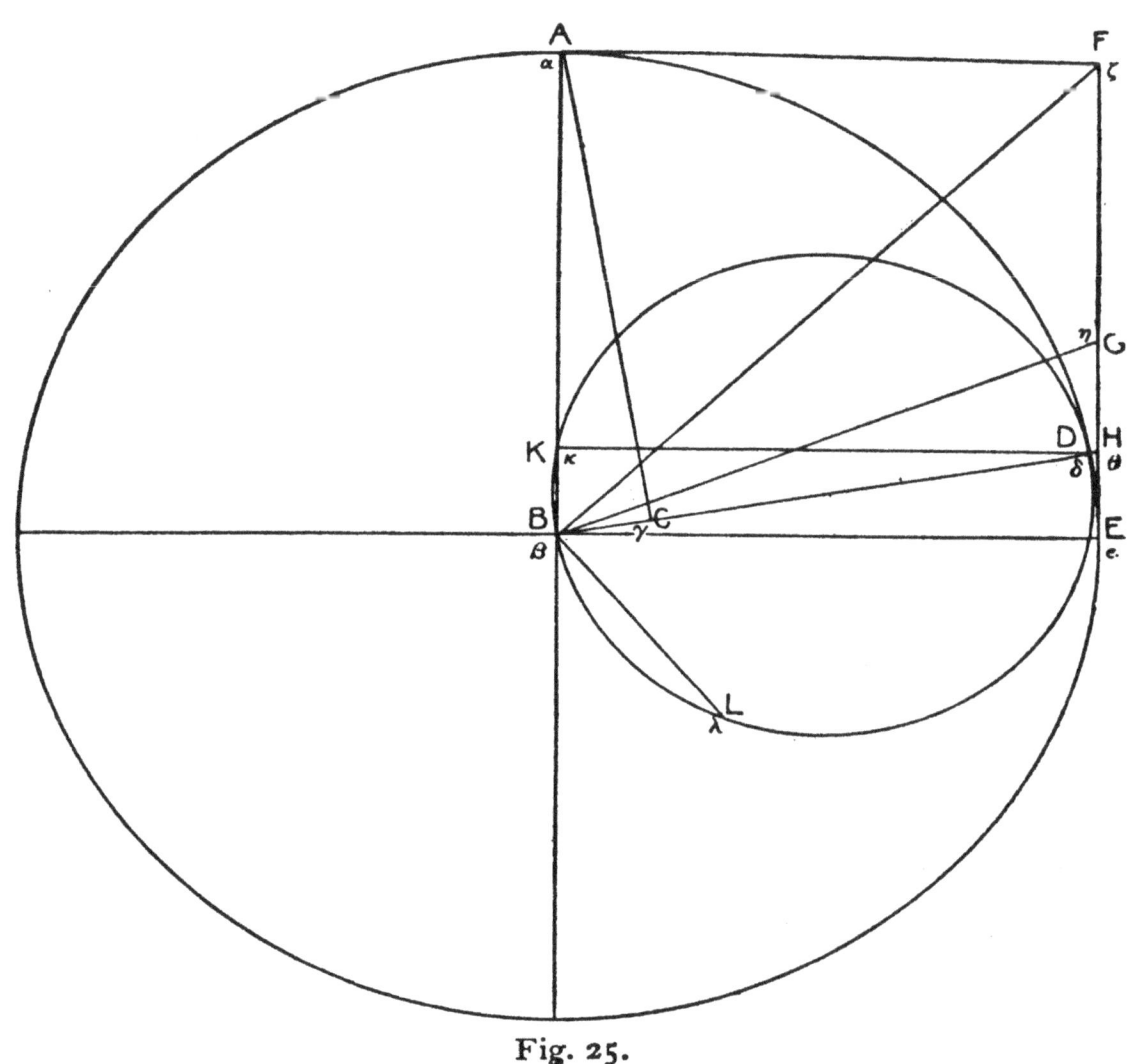

Fig. 25.

OF THE SUN AND MOON

therefore *GE* has to *EH* a ratio greater than that which 15 has to 2.

Next, since *BE* is equal to *EF*, and the angle at *E* is right, therefore the square on *FB* is double of the square on *BE*.

But, as the square on *FB* is to the square on *BE*, so is the square on *FG* to the square on *GE*;

therefore the square on *FG* is double of the square on *GE*.

Now 49 is less than double[1] of 25, so that the square on *FG* has to the square on *GE* a ratio greater than that which 49 has to 25;

therefore *FG* also has to *GE* a ratio greater than that which 7 has to 5.

Therefore, *componendo*, *FE* has to *EG* a ratio greater than that which 12 has to 5, that is, than that which 36 has to 15.

and let it meet *BH* in *H*. Let a circle be described with *B* as centre and *BH* as radius, meeting *BG* in *P* and *BE* produced in *Q*.

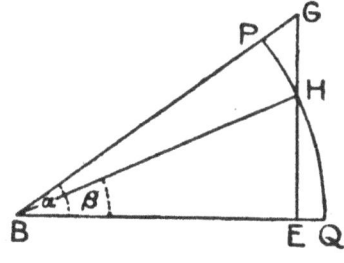

Fig. 26.

Then △ *GBH* : △ *HBE* > (sector *PBH*) : (sector *HBQ*);
therefore *GH* : *HE* > ∠ *GBH* : ∠ *HBE*,
and, *componendo*, *GE* : *HE* > ∠ *GBE* : ∠ *HBE*.

[1] Aristarchus here uses the well-known Pythagorean approximation to $\sqrt{2}$, namely $\frac{7}{5}$, one of the first of the successive approximations obtained by the development of the system of 'side-' and 'diagonal-' numbers (as to which see Theon of Smyrna, pp. 43, 44, ed. Hiller, and Proclus, *Comm. in Platonis rempublicam*, ed. Kroll, vol. ii, pp. 24, 25, 27–9, 393–400). The approximation $\frac{7}{5}$ is alluded to by Plato in the *Republic*, 546 C. Plato there speaks of the diagonal of the square, the side of which contains 5 units, and contrasts the 'irrational diameter of 5' (ἄρρητος διάμετρος τῆς πεμπάδος), which is of course $\sqrt{(50)}$, with the 'rational diameter' (ῥητὴ διάμετρος), which is the square root of 50 less a single unit, i.e. the square root of 49.

But it was also proved that GE has to EH a ratio greater than that which 15 has to 2;
therefore, *ex aequali*, FE has to EH a ratio greater than that which 36 has to 2, that is, than that which 18 has to 1;
therefore FE is greater than 18 times EH.

And FE is equal to BE;
therefore BE is also greater than 18 times EH;
therefore BH is much greater than 18 times HE.

But, as BH is to HE, so is AB to BC, because of the similarity of the triangles;
therefore AB is also greater than 18 times BC.

And AB is the distance of the sun from the earth, while CB is the distance of the moon from the earth; therefore the distance of the sun from the earth is greater than 18 times the distance of the moon from the earth.

Again, I say that it is also less than 20 times that distance.

For let DK be drawn through D parallel to EB, and about the triangle DKB let the circle DKB be described; then DB will be its diameter, because the angle at K is right.

Let BL, the side of a hexagon, be fitted into the circle.

Then, since the angle DBE is 1/30th of a right angle, the angle BDK is also 1/30th of a right angle;
therefore the circumference BK is 1/60th of the whole circle.

But BL is also one sixth part of the whole circle.

Therefore the circumference BL is ten times the circumference BK.

And the circumference BL has to the circumference BK a ratio greater than that which the straight line BL has to the straight line BK;[1]
therefore the straight line BL is less than ten times the straight line BK.

And BD is double of BL;
therefore BD is less than 20 times BK.

But, as BD is to BK, so is AB to BC;
therefore AB is also less than 20 times BC.

And AB is the distance of the sun from the earth, while BC is the distance of the moon from the earth; therefore the distance of the sun from the earth is less than 20 times the distance of the moon from the earth.

And it was before proved that it is greater than 18 times that distance.

---

[1] By the proposition proved in Ptolemy's *Syntaxis*, i, 10, pp. 43-4, ed. Heiberg. See, for his proof, the note on pp. 368-9, above.

OF THE SUN AND MOON

PROPOSITION 8.

When the sun is totally eclipsed, the sun and the moon are then comprehended by one and the same cone which has its vertex at our eye.

For since, if the sun is eclipsed, it is eclipsed because the moon is in front of it,

the sun must fall into the cone comprehending the moon and having its vertex at our eye.

And, if it falls into it, either it will exactly fit into it, or it must overlap it or fall short of it.

If now it should overlap it, the sun would not be totally eclipsed, but the portion which overlaps would be unobstructed.[1]

If, however, it should fall short, the sun would remain eclipsed for the time which it takes to pass through the portion by which it falls short.

But it is in fact totally eclipsed and does not remain eclipsed: for this is manifest from observation.[2]

Hence it can neither overlap nor fall short;

therefore it will exactly fit into the cone and will be comprehended by the cone comprehending the moon and having its vertex at our eye.

PROPOSITION 9.

The diameter of the sun is greater than 18 times, but less than 20 times, the diameter of the moon.

For let our eye be at A, let B be the centre of the sun, and C the centre of the moon when the cone comprehending both the sun and the moon has its vertex at our eye, that is, when the points A, C, B are in a straight line.

Let a plane be carried through ACB;

this plane will cut the spheres in great circles and the cone in straight lines.

---

[1] Gr. παραλλάττοι. As in Euclid, i. 7 and iii. 24, παραλλάττειν means to 'fall beside' or 'awry', to 'miss', to "pass by without touching'.

[2] It is evident from this that Aristarchus had not observed the phenomenon of an *annular* eclipse of the sun. The first mention of annular eclipses on record appears to be that quoted by Simplicius (on *De caelo*, ii. 12, p. 505, 7-9, Heiberg) from Sosigenes, the teacher of Alexander Aphrodisiensis (end of second century A.D.).

384 ON THE SIZES AND DISTANCES

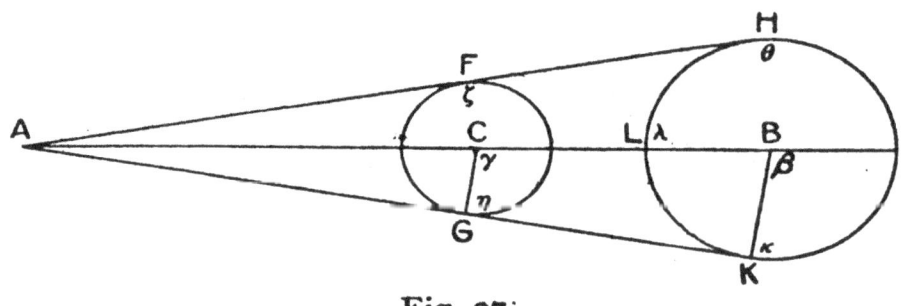

Fig. 27.

OF THE SUN AND MOON

Let it cut the spheres in the great circles *FG*, *KLH*, and the cone in the straight lines *AFH*, *AGK*,
and let *CG*, *BK* be joined.

Then, as *BA* is to *AC*, so will *BK* be to *CG*.

But it was proved that *BA* is greater than 18 times, but less than 20 times, *AC*. [Prop. 7]

Therefore *BK* is also greater than 18 times, but less than 20 times, *CG*.

PROPOSITION 10.

The sun has to the moon a ratio greater than that which 5832 has to 1, but less than that which 8000 has to 1.

Let *A* be the diameter of the sun, *B* that of the moon.

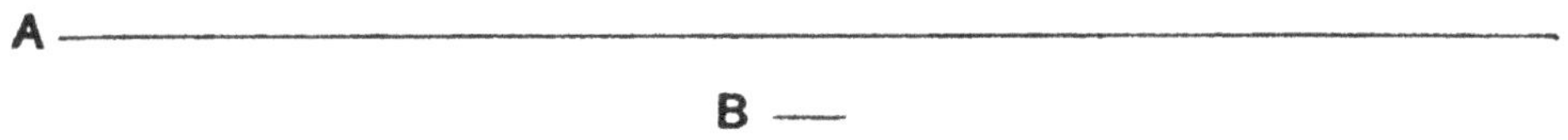

A ———————————————————————————

B ——

Fig. 28.

Then *A* has to *B* a ratio greater than that which 18 has to 1, but less than that which 20 has to 1.

Now, since the cube on *A* has to the cube on *B* the ratio triplicate of that which *A* has to *B*,
while the sphere about *A* as diameter also has to the sphere about *B* as diameter the ratio triplicate of that which *A* has to *B*,
therefore, as the sphere about *A* as diameter is to the sphere about *B* as diameter, so is the cube on *A* to the cube on *B*.

But the cube on *A* has to the cube on *B* a ratio greater than that which 5832 has to 1, but less than that which 8000 has to 1, since *A* has to *B* a ratio greater than that which 18 has to 1, but less than that which 20 has to 1.

Accordingly the sun has to the moon a ratio greater than that which 5832 has to 1, but less than that which 8000 has to 1.

PROPOSITION 11.

The diameter of the moon is less than 2/45ths, but greater than 1/30th, of the distance of the centre of the moon from our eye.

For let our eye be at A, and let B be the centre of the moon when the cone comprehending both the sun and the moon has its vertex at our eye.

I say that the above proposition is true.

Let AB be joined, and let the plane through AB be drawn; this plane will cut the sphere [i.e. the moon] in a circle and the cone in straight lines.

Let it cut the sphere in the circle CED and the cone in the straight lines AD, AC;
let BC be joined and produced to E.

Then it is manifest from what has before been proved [Prop. 4] that the angle BAC is 1/45th part of half a right angle;
and, in the same way as before, BC is less than 1/45th part of CA;
therefore BC is much less than 1/45th part of BA.

And CE is double of BC;
therefore CE is less than 2/45ths of AB.

Now CE is the diameter of the moon,
while BA is the distance of the centre of the moon from our eye.

Therefore the diameter of the moon is less than 2/45ths of the distance of the centre of the moon from our eye.

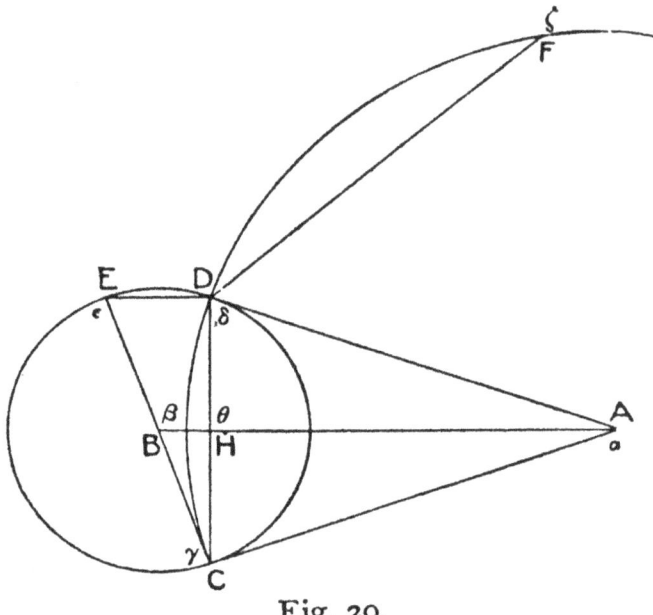

Fig. 29.

I say next that CE is also greater than 1/30th part of BA.

For let DE and DC be joined, and with centre A and distance AC let the circle CDF be described; let DF equal to AC be fitted into the circle CDF.

Then, since the right angle EDC is equal to the right angle BCA, while the angle BAC is also equal to the angle HCB,
therefore the remaining angle DEC is equal to the remaining angle HBC.

Therefore the triangle CDE is equiangular with the triangle ABC.

Therefore, as BA is to AC, so is EC to CD;
and, alternately, as AB is to CE, so is AC to CD, that is, DF to CD.

But again, since the angle DAC is 1/45th part of a right angle, the circumference CD is 1/180th part of the circle;
and the circumference DF is one sixth part of the whole circle;
thus the circumference CD is 1/30th part of the circumference DF.

And the circumference CD, being less than the circumference DF, has to the circumference DF itself a ratio less than that which the straight line CD has to the straight line FD.[1]

Therefore the straight line CD is greater than 1/30th of DF.

But FD is equal to AC;
therefore DC is greater than 1/30th of CA,
so that CE is also greater than 1/30th of BA [see above].

But it was before proved to be also less than 2/45ths.

PROPOSITION 12.

The diameter of the circle which divides the dark and the bright portions in the moon is less than the diameter of the moon, but has to it a ratio greater than that which 89 has to 90.

For let our eye be at A, and let B be the centre of the moon when the cone comprehending both the sun and the moon has its vertex at our eye;
let AB be joined, and let a plane be carried through AB; this plane will cut the sphere [i.e. the moon] in a circle and the cone in straight lines.

Let it cut the sphere in the circle DEC and the cone in the straight lines AD, AC, CD.

---

[1] For, by the proposition proved by Ptolemy (see note on Prop. 4, above),
$$FD : DC < (\text{arc } FD) : (\text{arc } DC),$$
and, by inversion, $(\text{arc } CD) : (\text{arc } DF) < CD : DF.$

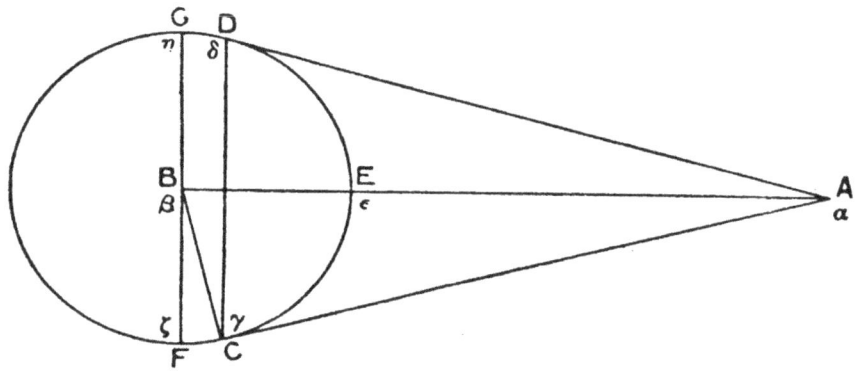

Fig. 30.

OF THE SUN AND MOON

Therefore CD is a diameter of the circle which divides the dark and the bright portions in the moon.

I say that CD is less than the diameter of the moon, but has to it a ratio greater than that which 89 has to 90.

Now, that CD is less than the diameter of the moon is manifest.

I say, then, that it also has to it a ratio greater than that which 89 has to 90.

For let FG be drawn through B parallel to CD, and let BC be joined.

Then again, in the same way as before, the angle DAC will be 1/45th part of a right angle,

and the angle BAC will be 1/90th part of a right angle;

but the angle BAC is equal to the angle CBF;

therefore the angle CBF is also 1/90th of a right angle,

that is, 1/90th of the angle FBE;

so that the circumference CF is also 1/90th of the circumference FCE.

Therefore the circumference CE has to the circumference ECF the ratio which 89 has to 90.

Now DEC is double of CE, and GEF double of ECF;

therefore the circumference DEC has to the circumference GEF the ratio which 89 has to 90.

And the straight line DC has to the straight line GF a ratio greater than that which the circumference DEC has to the circumference GEF.[1]

Therefore also the straight line DC has to the straight line GF a ratio greater than that which 89 has to 90.

---

[1] By the proposition quoted from Ptolemy, i. 10, pp. 43-4, ed. Heiberg. See note on Props. 4 and 11, above.

PROPOSITION 13.

The straight line subtending the portion intercepted within the earth's shadow of the circumference of the circle in which the extremities of the diameter of the circle dividing the dark and the bright portions in the moon move is less than double of the diameter of the moon, but has to it a ratio greater than that which 88 has to 45; and it is less than 1/9th part of the diameter of the sun, but has to it a ratio greater than that which 22 has to 225. But it has to the straight line drawn from the centre of the sun at right angles to the axis and meeting the sides of the cone a ratio greater than that which 979 has to 10125.

For let the centre of the sun be at A, let B be the centre of the earth, and C the centre of the moon when the eclipse first becomes total through the moon having fallen wholly within the earth's shadow.

Let a plane be carried through A, B, C;
this plane will cut the spheres in circles and the cone comprehending both the sun and the earth in straight lines.

Let it cut the spheres in the great circles DEF, GHK, LMN, the earth's shadow in the circle OLN in which the extremities of the diameter of the circle dividing the dark and the bright portions in the moon move, and the cone in the straight lines DGO, FKN.

Let ABL be the axis.

Then it is manifest that the axis ABL touches the circle LMN, because the shadow of the earth is of two moon-breadths, [Hyp. 5] that the circumference NLO is bisected by the axis ABL,
and further that the moon has for the first time fallen within the earth's shadow.

Let ON, NL, BN, LO be joined.

Therefore LN is the diameter of the circle dividing the dark and the bright portions in the moon.

And BN touches the circle $LNPM$,
because the point B is at our eye, and LN is the diameter of the circle dividing the dark and the bright portions in the moon.

ON THE SIZES AND DISTANCES

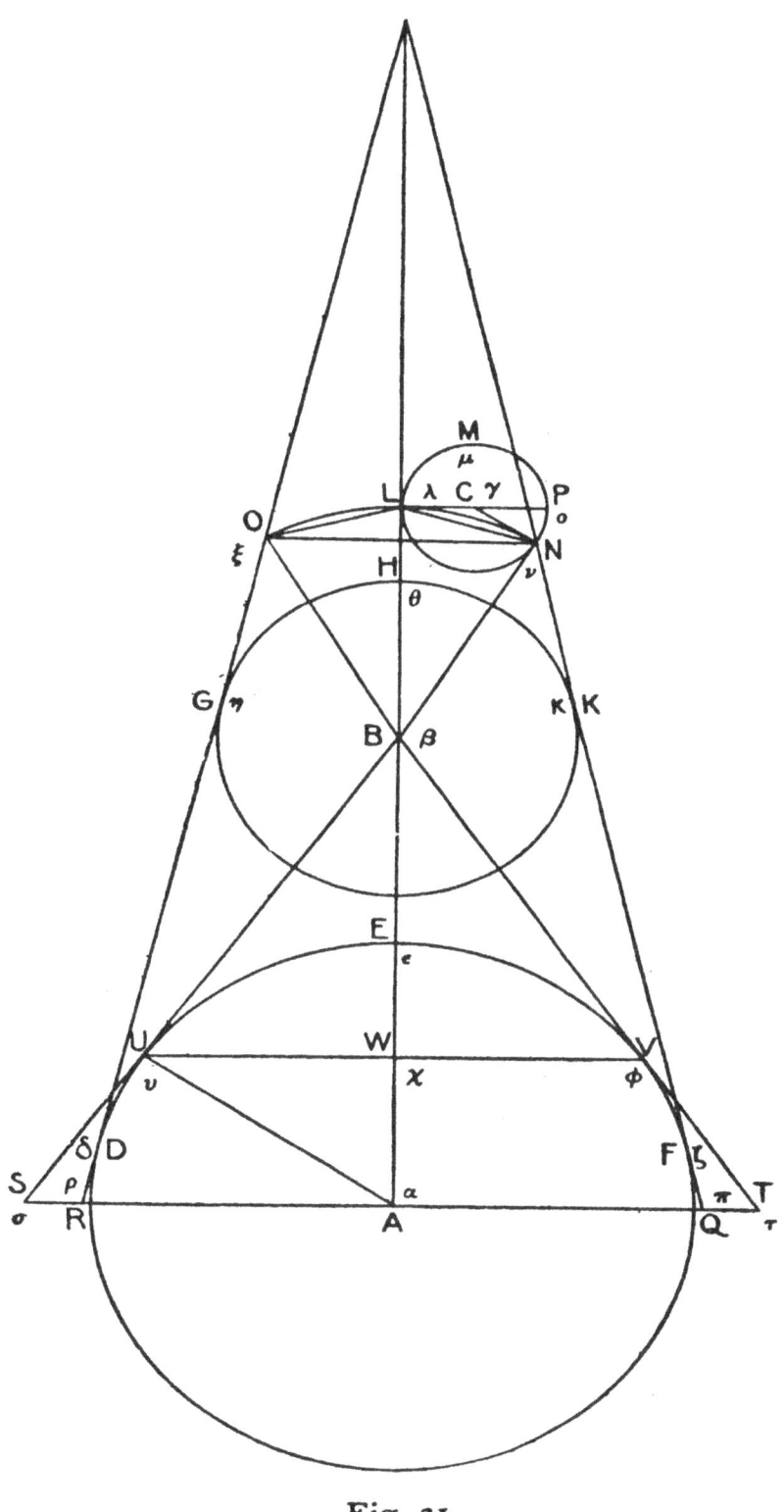

Fig. 31.

Now, since OL, LN are equal, their sum is double of LN, so that ON is less than double of LN.

Let LC, CN be joined; and let LC be carried through to P.

Therefore ON is much less than double of LP.

And, since CL is perpendicular to BL, therefore it is parallel to ON.

Therefore the angle LON is equal to the angle CLN.

And NL is equal to LO, and LC to CN; therefore the triangle ONL is similar to the triangle LNC; therefore, as ON is to NL, so is NL to LC.

But NL has to LC a ratio greater than that which 89 has to 45;[1] that is, the square on NL has to the square on LC a ratio greater than that which 7921 has to 2025.

Therefore the square on ON also has to the square on NL a ratio greater than that which 7921 has to 2025, and (therefore) ON has to LP a ratio greater than that which 7921 has to 4050.[2]

---

[1] For $NL : LP > 89 : 90$, by the preceding proposition.

[2] We have $ON : NL = NL : LC$;

therefore $ON : LC = $ (sq. on ON) : (sq. on NL)

$> 7921 : 2025$,

whence $ON : LP > 7921 : 4050$.

OF THE SUN AND MOON

But 7921 also has to 4050 a ratio greater than that which 88 has to 45;[1]

therefore NO has to LP a ratio greater than that which 88 has to 45.

Therefore the straight line which subtends the portion intercepted within the earth's shadow of the circumference of the circle in which the extremities of the diameter of the circle dividing the dark and the bright portions in the moon move is less than double of the diameter of the moon, but has to it a ratio greater than that which 88 has to 45.

The same suppositions being made, let QAR be drawn from A at right angles to AB.

I say that ON is less than 1/9th part of the diameter of the sun, but has to it a ratio greater than that which 22 has to 225, and has to QR a ratio greater than that which 979 has to 10125.

For, since it was proved that ON is less than double of the diameter of the moon,

while the diameter of the moon is less than 1/18th part of the diameter of the sun, [Prop. 9]

therefore ON is less than 1/9th part of the diameter of the sun.

Again, since ON has to the diameter of the moon a ratio greater than that which 88 has to 45,

while the diameter of the moon has to the diameter of the sun a ratio greater than that which 45 has to 900:

for, since the diameter of the moon has to the diameter of the sun a ratio greater than that which 1 has to 20, we have only to multiply throughout by 45:

therefore (*ex aequali*) ON has to the diameter of the sun a ratio greater than that which 88 has to 900, that is, than that which 22 has to 225.

Now let BUS, BVT be drawn from B touching the circle DE: and let UV, UA be joined.

Then, as the diameter of the circle dividing the dark and the bright portions in the moon is to the diameter of the moon, so is UV to the diameter of the sun, because the sun and the moon are

---

[1] If we develop $\frac{7921}{4050}$ as a continued fraction, we easily obtain the approximation $1+\frac{1}{1+}\frac{1}{21+}\frac{1}{2}$, which is in fact $\frac{88}{45}$. See the similar case in Prop. 15, p. 407, and the observation thereon, p. 336 *ad fin.*

comprehended by one and the same cone having its vertex at our eye.[1]

But the diameter of the circle dividing the dark and the bright portions in the moon has to the diameter of the moon a ratio greater than that which 89 has to 90;   [Prop. 12]
therefore UV also has to the diameter of the sun a ratio greater than that which 89 has to 90.

Therefore WU also has to UA a ratio greater than that which 89 has to 90.

But, as WU is to UA, so is UA to AS, because SA, UW are parallel;
therefore UA also has to AS a ratio greater than that which 89 has to 90;
therefore UA has to AR a ratio much greater than that which 89 has to 90.

The same is true of the doubles;
therefore the diameter of the sun has to QR a ratio greater than that which 89 has to 90.

But it was proved [above] that ON has to the diameter of the sun a ratio greater than that which 22 has to 225;
therefore, *ex aequali*, ON has to QR a ratio much greater than that which the product of 22 and 89 has to the product of 90 and 225, that is, 1958 to 20250, or, if the halves be taken, 979 to 10125.

Proposition 14.

The straight line joined from the centre of the earth to the centre of the moon has to the straight line cut off from the axis towards the centre of the moon by the straight line subtending the (circumference) within the earth's shadow a ratio greater than that which 675 has to 1.

For let the same figure be drawn as before;
and let the moon be so placed that its centre is on the axis of the cone comprehending both the sun and the earth;

---

[1] The proof, which is given by Commandinus, is obvious. The fact cannot be seen from our figure, which, owing to exigency of space, could not be drawn so as to make the angles LBN, UBV equal.

400 ON THE SIZES AND DISTANCES

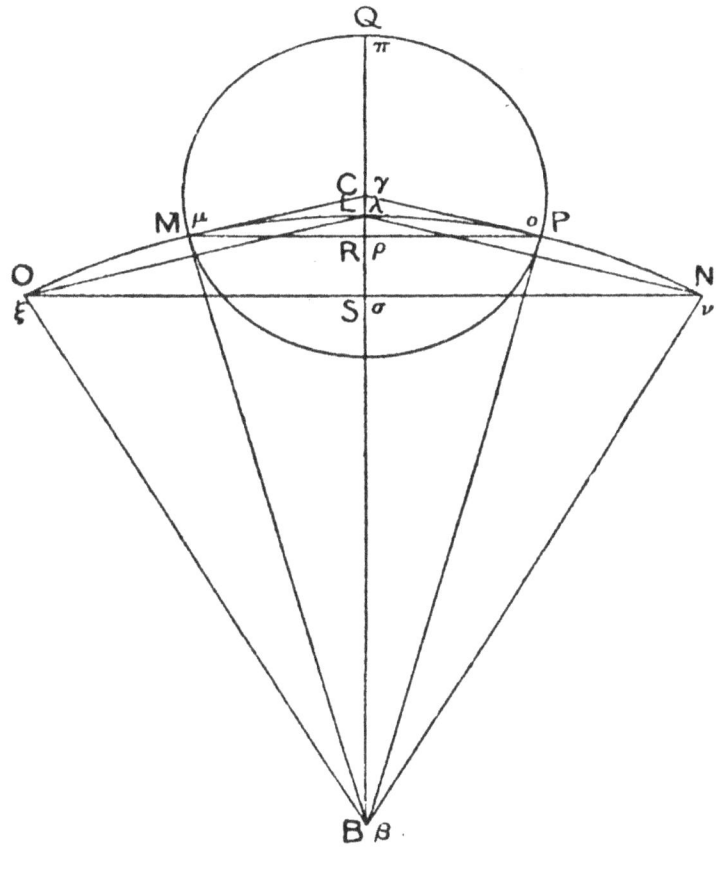

Fig. 32.

OF THE SUN AND MOON

let its centre be C, and let the great circle QPM in the sphere [i.e. the moon] be in the same plane with the rest of the figure.[1]

Let MP be joined;

therefore MP is a diameter of the circle which divides the dark and the bright portions in the moon.

Let MB, BP, LO, OB, MC be joined.

Therefore MB, BP touch the circle MPQ,

because PM is a diameter of the circle which divides the dark and the bright portions in the moon.

And, since OL is equal to MP—for each of them is a diameter of the circle which divides the dark and the bright portions in the moon—

therefore the circumference OML is equal to the circumference MLP;

therefore OM is also equal to LP.

But LP is equal to LM;

therefore OM is also equal to LM.

And OB is equal to BL,

because the point B is the centre of the earth, and the earth has

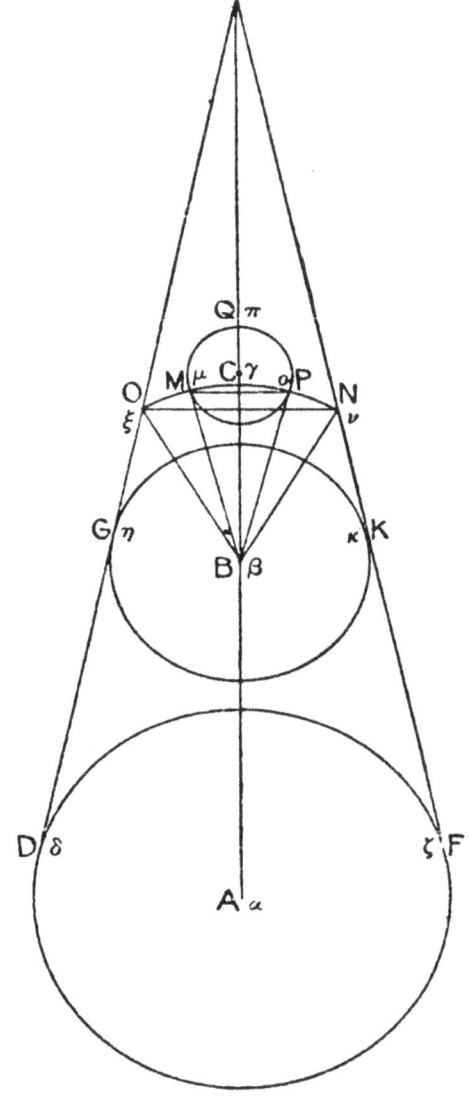

Fig. 33.

---

[1] Literally 'in the same plane with *them*' (αὐτοῖς), which no doubt means the axis and the sections of the sun and moon made by the plane originally assumed, which also contains the circle in which the diameter of the 'dividing circle' in the moon moves while the moon is passing through the earth's shadow.

the relation of a point and centre to the sphere in which the moon moves [Hyp. 2], while the circle MPQ is in the same plane;
therefore BM is perpendicular to OL.

But CM is also perpendicular to BM;
therefore CM is parallel to OL.

And SO is also parallel to MR;
therefore the triangle LOS is similar to the triangle MRC.

Therefore, as SO is to MR, so is SL to RC.

But SO is less than double of MR,
since ON is also less than double of MP;   [Prop. 13]
therefore SL is also less than double of CR,
so that SR is much less than double of RC.

Therefore SC is less than triple of CR;
therefore CR has to CS a ratio greater than that which 1 has to 3.

And since, as BC is to CM, so is CM to CR,
while BC has to CM a ratio greater than that which 45 has to 1,
    [see Prop. 11]
therefore CM also has to CR a ratio greater than that which 45 has to 1.

But CR also has to CS a ratio greater than that which 1 has to 3;
therefore, *ex aequali*, CM has to CS a ratio greater than that which 45 has to 3, that is, than that which 15 has to 1.

And it was proved that BC has to CM a ratio greater than that which 45 has to 1;
therefore, *ex aequali*, BC has to CS a ratio greater than that which 675 has to 1.

PROPOSITION 15.

The diameter of the sun has to the diameter of the earth a ratio greater than that which 19 has to 3, but less than that which 43 has to 6.

For let A be the centre of the sun, B the centre of the earth, C the centre of the moon when the eclipse is total, so as to secure that A, B, C may be in a straight line.

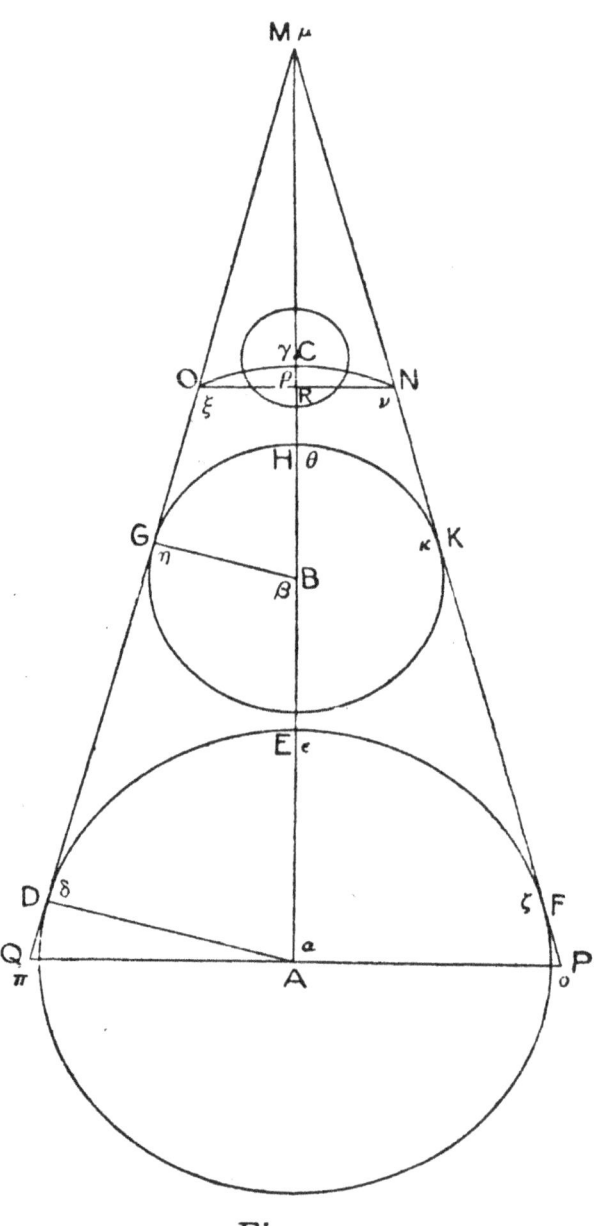

Fig. 34.

Let a plane be carried through the axis, and let it cut the sun in the circle *DEF*, the earth in *GHK*, the shadow in the circumference *NO*, and the cone in the straight lines *DM*, *FM*.

Let *NO* be joined, and from *A* let *PAQ* be drawn at right angles to *AM*.

Then, since *NO* is less than 1/9th part of the diameter of the sun, [Prop. 13]
therefore *PQ* has to *NO* a ratio much greater than that which 9 has to 1.

Therefore *AM* also has to *MR* a ratio greater than that which 9 has to 1;
and, *convertendo*, *MA* has to *AR* a ratio less than that which 9 has to 8.

Again, since *AB* is greater than 18 times *BC*, [Prop. 7]
therefore it is much greater than 18 times *BR*;
therefore *AB* has to *BR* a ratio greater than that which 18 has to 1;
therefore, inversely, *BR* has to *BA* a ratio less than that which 1 has to 18;
therefore, *componendo*, *RA* has to *AB* a ratio less than that which 19 has to 18.

But it was proved that *MA* also has to *AR* a ratio less than that which 9 has to 8;
therefore, *ex aequali*, *MA* will have to *AB* a ratio less than that which 171 has to 144, and therefore less than that which 19 has to 16: for parts have the same ratio as the same multiples of them:

therefore, *convertendo*, AM has to BM a ratio greater than that which 19 has to 3.

But, as AM is to MB, so is the diameter of the circle DEF to the diameter of the circle GHK;

therefore the diameter of the sun has to the diameter of the earth a ratio greater than that which 19 has to 3.

Again, I say that it has to it a ratio less than that which 43 has to 6.

For, since BC has to CR a ratio greater than that which 675 has to 1, [Prop. 14]

therefore, *convertendo*, CB has to BR a ratio less than that which 675 has to 674.

But AB also has to BC a ratio less than that which 20 has to 1; [Prop. 7]

therefore, *ex aequali*, AB will have to BR a ratio less than that which 13500 has to 674, that is, than that which 6750 has to 337;

therefore, inversely and *componendo*, RA has to AB a ratio greater than that which 7087 has to 6750.

Now, since NO has to PQ a ratio greater than that which 979 has to 10125, [Prop. 13]

therefore, inversely, PQ has to NO a ratio less than that which 10125 has to 979.

And, as PQ is to NO, so is AM to MR;

therefore AM also has to MR a ratio less than that which 10125 has to 979;

therefore, *convertendo*, MA has to AR a ratio greater than that which 10125 has to 9146.

But RA also has to AB a ratio greater than that which 7087 has to 6750;

therefore, *ex aequali*, MA will have to AB a ratio greater than that which the number representing the product of 10125 and 7087 has to the number representing the product of 9146 and 6750, that is, 71755875 to 61735500.

But 71755875 has to 61735500 a ratio greater than that which 43 has to 37;[1]

therefore MA also has to AB a ratio greater than that which 43 has to 37;

---

[1] As to this approximation see p. 336 *ad fin.*

therefore, *convertendo*, *AM* has to *MB* a ratio less than that which 43 has to 6.

But, as *AM* is to *BM*, so is the diameter of the sun to the diameter of the earth;

therefore the diameter of the sun has to the diameter of the earth a ratio less than that which 43 has to 6.

And it was before proved that it has to it a ratio greater than that which 19 has to 3.

PROPOSITION 16.

The sun has to the earth a ratio greater than that which 6859 has to 27, but less than that which 79507 has to 216.

For let *A* be the diameter of the sun, *B* that of the earth.

Now it is proved that, as the sphere of the sun is to the sphere of the earth, so is the cube on the diameter of the sun to the cube

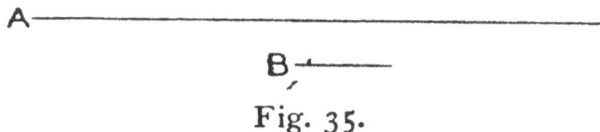

Fig. 35.

on the diameter of the earth, just as in the case of the moon [cf. Prop. 10].

Thus, since, as the cube on *A* is to the cube on *B*, so is the sun to the earth,
while the cube on *A* has to the cube on *B* a ratio greater than that which 6859 has to 27, but less than that which 79507 has to 216:
for *A* has to *B* a ratio greater than that which 19 has to 3, but less than that which 43 has to 6: [Prop. 15]
it follows that the sun has to the earth a ratio greater than that which 6859 has to 27, but less than that which 79507 has to 216.

PROPOSITION 17.

The diameter of the earth is to the diameter of the moon in a ratio greater than that which 108 has to 43, but less than that which 60 has to 19.

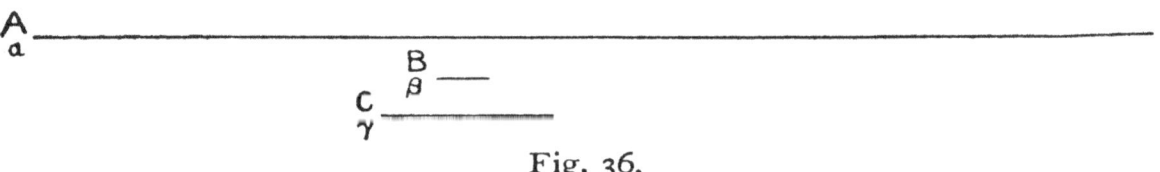

Fig. 36.

For let A be the diameter of the sun, B that of the moon, C that of the earth.

Then, since A has to C a ratio less than that which 43 has to 6, [Prop. 15]
therefore, inversely, C has to A a ratio greater than that which 6 has to 43.

But A also has to B a ratio greater than that which 18 has to 1; [Prop. 9]
therefore, *ex aequali*, C has to B a ratio greater than that which 108 has to 43.

Again, since A has to C a ratio greater than that which 19 has to 3, [Prop. 15]
therefore, inversely, C has to A a ratio less than that which 3 has to 19.

But A also has to B a ratio less than that which 20 has to 1; [Prop. 9]
therefore, *ex aequali*, C has to B a ratio less than that which 60 has to 19.

PROPOSITION 18.

The earth is to the moon in a ratio greater than that which 1259712 *has to* 79507, *but less than that which* 216000 *has to* 6859.

For let A be the diameter of the earth, B that of the moon; therefore A has to B a ratio greater than that which 108 has to 43, but less than that which 60 has to 19. [Prop. 17]

Therefore also the cube on A has to the cube on B a ratio

A ———————————
B ———
Fig. 37.

greater than that which 1259712 has to 79507, but less than that which 216000 has to 6859.

But, as the cube on A is to the cube on B, so is the earth to the moon;
therefore the earth has to the moon a ratio greater than that which 1259712 has to 79507, but less than that which 216000 has to 6859.

www.ingramcontent.com/pod-product-compliance
Lightning Source LLC
Chambersburg PA
CBHW081639220526
45468CB00009B/2495